BASIC BIOLOGY

Inheritance and Evolution

DENISE WALKER

A⁺
Smart Apple Media

First published in 2006 by Evans Brothers Ltd.
2A Portman Mansions, Chiltern Street,
London W1U 6NR

This edition published under license from
Evans Brothers Ltd. All rights reserved.
Copyright © 2006 Evans Brothers Ltd.

Series editor: Harriet Brown, Editor: Harriet
Brown, Design: Robert Walster, Illustrations:
Q2A Creative

Published in the United States by
Smart Apple Media
2140 Howard Drive West, North Mankato,
Minnesota 56003

U.S. publication copyright © 2007
Smart Apple Media
International copyright reserved in all
countries. No part of this book may be
reproduced in any form without written
permission from the publisher.
Printed in China

Library of Congress Cataloging-in-
Publication Data

Walker, Denise.
Inheritance and evolution /
by Denise Walker.
p. cm. — (Basic biology)
Includes index.
ISBN-13: 978-1-58340-989-3
1. Human genetics. 2. Human evolution.
I. Title.

QH431.W285 2006
576—dc22   2006000346

9 8 7 6 5 4 3 2 1

Contents

Introduction

The variety of life on planet Earth is astounding. Life inhabits almost every corner of the globe, from the hot, dry deserts to the

harsh, icy poles and everywhere in between. For millions of years, species of plants, animals, and microscopic organisms have been adapting their lifestyles to suit their habitats. They have evolved into new species better suited to the environment. Species unable to adapt eventually die out and become extinct. In fact, around 99.9 percent of all organisms that ever existed are now extinct.

In evolutionary terms, humans have existed on planet Earth for only a blink of an eye, but we are very successful. We have evolved from our primitive beginnings into intelligent beings who now dominate the globe.

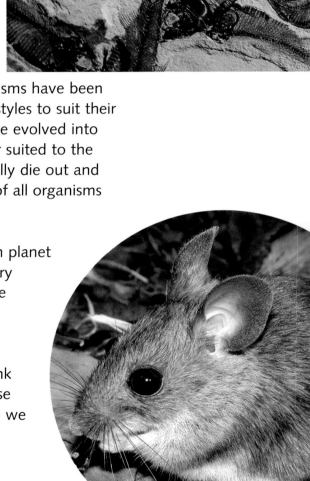

Each human being is unique. We look and think differently from one another, and we pass these characteristics on to the next generation when we reproduce. This is inheritance, and through inheritance, we help to create the variation in our species that makes the world such an interesting place to live.

This book guides you through inheritance and explains the fascinating ideas behind natural selection and evolution. It looks at the effects humans can have on evolution and examines how scientists classify and identify species. Feature boxes will help you unravel more about the mysteries of inheritance and evolution.

Did you know?

▶ Look for these boxes—they contain surprising and fascinating facts about inheritance, evolution, and classification.

Test yourself

▶ Use these boxes to see how much you've learned. Try to answer the questions without looking at the book, but take a look if you are really stuck.

Investigate

▶ These boxes contain experiments you can carry out at home. The equipment you will need is usually cheap and easy to find.

Time travel

These boxes contain scientific discoveries from the past and fascinating developments that pave the way for the advance of science in the future.

Answers

At the end of this book, on pages 46 and 47, you will find the answers to questions from the "Test yourself" and "Investigate" boxes.

Glossary

Words highlighted in **bold** are described in detail in the glossary on pages 46 and 47.

What is variation?

In a group of people, such as your class at school or your family, there are similarities and differences between individuals. Each person is a human being with two eyes, two ears, and two legs. However, it is our differences, or variations, such as our size, shape, and intelligence, that make every one of us unique. Variations are found within any group of animals or plants.

VARIATION

Organisms from the same **species** can reproduce together. Humans are one species, and cats are another species. Characteristics such as skin, hair, and eye color, and height and weight, vary a great deal among humans—this is our variation. Variation refers to the differences between individuals or groups of individuals.

Organisms from the same species have less variation than organisms from different species. In other words, two humans have more in common than a human and a cat. There are two main types of variation: **continuous variation** and **discontinuous variation**.

▼ Human beings display a wide range of variations.

CONTINUOUS VARIATION

Continuous variation occurs when there is a gradual change in a particular feature within a group. For example, look at the people in your class and notice the height differences among everyone. A few class members are likely to be very tall, while others are very short. However, most of the class falls somewhere between the two extremes. This pattern is repeated with lots of other characteristics, such as foot size and body weight.

If the heights of a group of people of the same age and sex are plotted on a graph, a distinctive curve is produced. This is called a **normal distribution curve**. It shows that a few people are very tall or very short, but the heights of most people fall in the middle. There is a gradual change in height from the shortest to the tallest within the group— this is continuous variation.

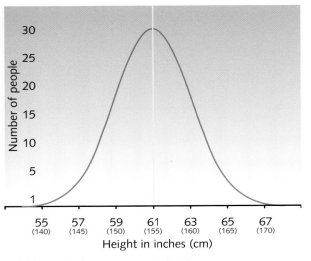

▲ This graph shows a normal distribution curve.

DISCONTINUOUS VARIATION

Discontinuous variation occurs when an individual can be put into one category or another depending on whether or not he or she has a particular characteristic. For example, the human hairline can have a distinctive peak, called a "widow's peak." Individuals either have this characteristic or they do not. We can plot this type of information as a bar chart. Blood group type is another example of discontinuous variation. There are four different blood groups, and all human beings belong to one of them.

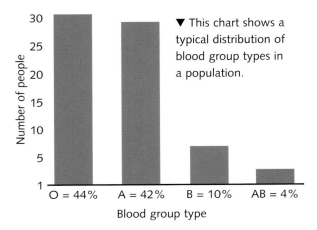

▼ This chart shows a typical distribution of blood group types in a population.

Continuous and discontinuous variations are found throughout nature and can usually be identified by the following rules:
(1) Continuous variation involves features that can be measured, such as height.
(2) Discontinuous variation involves features that cannot be measured, such as hairline, blood group, and gender.

INVESTIGATE

▶ Do a survey of your class to find out the shoe size of all individuals.
▶ Plot this information on a graph to see if there is a normal distribution curve.

TEST YOURSELF

▶ Write down two more examples of continuous variation.
▶ Write down two more examples of discontinuous variation.

CANINES

Variation does not only occur in human populations. Differences between individuals of the same species exist throughout the animal and plant kingdoms. Consider these two animals (right and below right). What do they have in common? How are they different from one another?

One dog (above right) is a dachshund and has a long back, short legs, and short fur. The other dog (below right) is a spaniel and is easily recognized by its unusually long ears and long fur. Both animals are dogs, but they have very different characteristics. So what is it that makes these animals dogs?

Animals within a species produce offspring that can also breed together. If spaniels and dachshunds breed, they produce offspring that is also capable of breeding. Therefore, spaniels and dachshunds belong to the same species. Animals from different species cannot breed together.

The breeding of dogs with different variations occurs regularly, which is how we end up with mongrels. Mongrels are a combination of different dog types. Any two dogs can

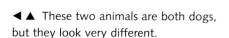

◀ ▲ These two animals are both dogs, but they look very different.

breed with one another as long as one is not huge and the other tiny. Humans have taken advantage of this and have intentionally bred dogs to produce particular variations, such as dogs with long hair, the ability to run fast, or a friendly personality.

ROSES

It is not only animals that exhibit variation within a species. Plant species, when naturally crossbred in the wild, or when deliberately crossbred by humans, show variation. For example, the variations of a rose species include flower color, speed of growth, and size. Look at the photographs of the two roses. What do they have in common? What are their differences?

The two plants can breed with each other to produce offspring.

◀ This rose has large, yellow petals.

8

The offspring could have characteristics from one of the parent plants or both. Gardeners crossbreed plants to achieve particular variations. For example, if a gardener wants to produce a large rose bush with a yellow flower, he would choose one large bush and one with yellow flowers and breed them; the offspring are called hybrids. It is likely that one of the hybrids would be large and have yellow flowers.

Animals and plants within a species can be naturally or artificially varied. However, they must retain enough characteristics in common to allow the species to reproduce and create future generations. If there are so many changes in a population that individuals from the new and the old populations could not breed together to create future generations, a new species has been created (see pages 24–25).

▲ This rose looks different from the flower on the opposite page, yet they are both roses.

DID YOU KNOW?

▶ A horse can breed with a donkey to produce a mule (right). Mules cannot breed.
▶ In the 1960s, Michigan State University biologist Guy Bush studied differences in fruit fly populations. The fruit fly population originally lived, fed, and mated on the fruit of hawthorn trees. However, some flies from the population developed a taste for apples and moved to a nearby apple tree. Bush was convinced that the two groups could no longer breed. In the 1980s, researchers identified differences between the two populations, which showed that the groups had not been mating for many generations. They had become different species (see page 25).

Genetic variation and inheritance

Individuals within a species can be highly varied. They are classified as the same species if they are still able to mate with one another to produce offspring who are also able to reproduce. But why do individuals look different from one another? Animals and plants inherit features from their parents, which affect the way they look.

GENES

The human body is made of millions of **cells**. Inside each cell is a nucleus. The nucleus contains **genes**, which carry the information that describes the entire structure of the human body. The genes are a combination of information from the father and mother. Inheritance refers to the genes gained from the parents when a new individual is created.

▲ These siblings have inherited their features from their mother and father.

HOW ARE GENES INHERITED?

During reproduction, one egg cell from the mother joins together with one sperm cell from the father to make a new human being. These sex cells contain only one copy of each gene. The rest of the cells in the body contain two copies of each gene. Variants of genes are called **alleles**. For example, when eye color is inherited, the offspring gains one eye color allele from the mother and one eye color allele from the father. Together, the two alleles determine the eye color in the offspring.

GENES DESCRIBE FEATURES

Snails belonging to the species *Cepaea nemoralis* have either plain shells or banded shells. There are two alleles that describe this feature. One describes a plain shell, and the other describes a banded shell. The allele describing a plain shell is dominant over the

◄ This snail has at least one "plain shell" allele.

▶ This dog (right) has one blue eye and one brown eye. Many animals, including humans, can have one brown and one blue eye. Actress Kate Bosworth and musician David Bowie have this condition, called heterochromia iridium. It happens when simple genetic patterns are not followed and there is no dominance. Instead, a mixture of alleles or a mutation in an allele for eye color can result in an individual having two different-colored eyes. Interestingly, such individuals often have hearing problems, which suggests that genes also cause the hearing loss.

▶ The vast majority of the world's population has dark eyes, ranging from light brown to nearly black. Dark eye color genes are dominant over paler eye color genes.

allele for a banded shell. The banded allele is therefore recessive. If both alleles are present in a snail, the plain shell is the characteristic that is displayed and is called the **phenotype**. The genetic makeup of the snail is called its **genotype**. The combination of alleles in an individual causes its genetic variation.

▶ This snail has two recessive "banded shell" alleles.

MUTATIONS CAUSE VARIATIONS

When a gene changes its form suddenly, it will not act as it was supposed to and can give rise to unexpected features. The sudden alteration of a gene is called a **mutation**. Mutations happen by chance in nature and can be caused by exposure to radiation and certain chemicals. Cancers are usually caused by genetic mutations that cause cells to reproduce uncontrollably. Mutations are often harmful, as in cancers, but some mutations have no effect at all or, in rare cases, a beneficial effect. For example, natural mutations have produced particularly woolly sheep. Man has taken advantage of this and has bred these woolly offspring to increase wool production.

Environmental variation

Genes determine which characteristics plants and animals inherit. However, the surroundings and lifestyle of the individual affect how the inherited characteristics develop. This is called environmental variation. Environmental variation can be caused by factors such as climate, water availability, diet, culture, and physical accidents.

ENVIRONMENTAL VARIATION IN PLANTS

Imagine you have a crop of vegetables to grow. To test how the environment affects the plants, you split the crop and plant parts of it in two different places. The first crop receives plenty of water, sunlight, and nutrient-rich soil. These plants grow well and produce large, healthy vegetables. The second crop does not receive enough water or sunlight and is grown in poor soil. These plants are deprived of important environmental factors and produce only small vegetables, or none at all. Even though the plants are genetically very similar, environmental variation results in two very different crops.

ENVIRONMENTAL VARIATION IN TWINS

The environment does not only affect plants; it also affects animals, including humans. One way to study the effects of the environment is to examine identical twins. Identical twins have exactly the same genetic information. They grow from a single fertilized egg (one egg that has been fertilized by one sperm), which splits into two individuals.

◄ To grow a high-quality crop of vegetables, the environmental conditions must be correct.

Most identical twins are brought up in the same environment and are exposed to the same events and experiences. However, when identical twins are separated at birth, differences between them are due to environmental factors. For example, if one twin is reared in a healthy environment with a good diet and plenty of exercise, and the other is reared in a less healthy environment with a lack of food and very little exercise, there is likely to be a clear difference in adult body shape.

▲ Identical twins have exactly the same genes.

AGING—ENVIRONMENT OR GENES?

In 1986, the Minnesota Twin Study of Adult Development began to identify what causes individual differences in aging. They discovered that as twins became older, their genes caused changes in their personality and activity levels. However, those who chose an active lifestyle and continued to use their brains remained more alert and had stronger feelings of well-being. Scientists believe that our genes determine many of our features, but the environment also plays an important role in defining who we are.

ENVIRONMENTAL VARIATION OVER TIME

Better conditions can lead to environmental variations over a period of time. Over the last 170 years, the average age of menstruation in females has lowered from 16.5 years to 12 years. Scientists believe this is because living conditions and diets are better. Menstruation in females begins when an individual reaches a certain weight. As conditions have improved, that weight is reached sooner, so the age of menstruation has decreased.

Adult height has increased since World War II (1939–1945). This is particularly spectacular in Japan. During the years following the war, living conditions for young people improved, which led to a rapid increase in the height of the population over a fairly short time.

DID YOU KNOW?

▶ According to a study of 522 pairs of identical and fraternal twins, optimism (a positive outlook) and pessimism (a negative outlook) are influenced by genes. However, the environment influences only optimism, and not pessimism. This means that the environment and genetics contribute to an optimistic outlook, but pessimism is largely controlled by genes.

▶ Not all identical twins are truly identical. In one pair of identical twins, one twin was healthy and a gymnast, but the other had a genetic condition called muscular dystrophy and died by the age of 16. This can happen when twins differ in the way they shut off certain **chromosomes**. This happens only with female identical twins.

Mendel and inheritance

Much of what we know about genes and inheritance has been discovered in the last 50 years. But before this time, there was much important work that provided scientists with a foundation on which to build new discoveries. A key figure in the science of inheritance was Gregor Mendel, an Austrian monk and teacher.

MENDEL AND HIS PEAS

Mendel was born in 1822 and lived in a monastery. There he was responsible for a small garden plot, where he was able to carry out some simple experiments on pea plants. Mendel deliberately chose pea plants because:

(1) They fertilized quickly, so he didn't have to wait too long for his results.

▼ Gregor Mendel—the first geneticist.

(2) Male and female parts were present on the same plant.

(3) The plants were easy to keep.

(4) Pure lines were available. (Pure lines are plants that always produce offspring with the same features as the parent plant. For example, if two tall plants are crossbred, all the offspring plants will be tall.)

MENDEL'S FIRST EXPERIMENTS

Mendel began to cross his pea plants and came up with some very clear observations that paved the way for understanding inheritance patterns. Mendel crossed pure "round pea plants" with pure "wrinkled pea plants." No matter how many times he did this, all of the offspring had round peas, and none of them had wrinkled peas. From this, Mendel concluded that information could be passed on from parents to offspring. Today, we know that this information is passed on through our genes.

Mendel also noticed that the pea plants were not a mixture of the characteristics of the parent plants. In other words, none of the plants had a mixture of round and wrinkled peas. When two characteristics were involved, one was always stronger than the other. In this case, the gene for round peas was stronger than the gene for wrinkled peas.

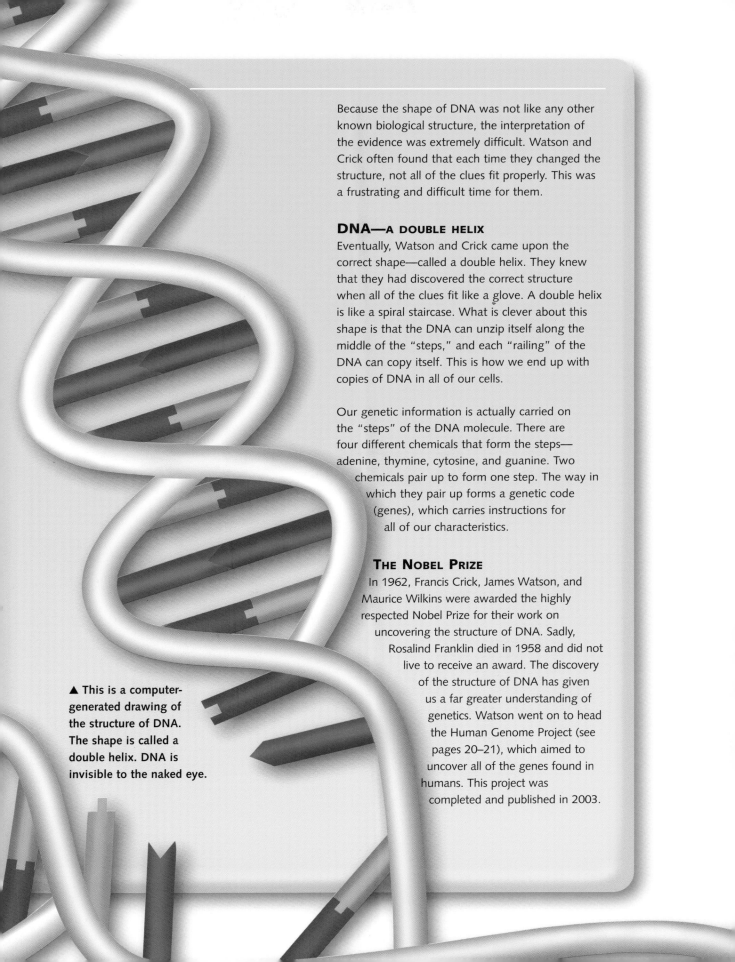

Because the shape of DNA was not like any other known biological structure, the interpretation of the evidence was extremely difficult. Watson and Crick often found that each time they changed the structure, not all of the clues fit properly. This was a frustrating and difficult time for them.

DNA—A DOUBLE HELIX

Eventually, Watson and Crick came upon the correct shape—called a double helix. They knew that they had discovered the correct structure when all of the clues fit like a glove. A double helix is like a spiral staircase. What is clever about this shape is that the DNA can unzip itself along the middle of the "steps," and each "railing" of the DNA can copy itself. This is how we end up with copies of DNA in all of our cells.

Our genetic information is actually carried on the "steps" of the DNA molecule. There are four different chemicals that form the steps—adenine, thymine, cytosine, and guanine. Two chemicals pair up to form one step. The way in which they pair up forms a genetic code (genes), which carries instructions for all of our characteristics.

THE NOBEL PRIZE

In 1962, Francis Crick, James Watson, and Maurice Wilkins were awarded the highly respected Nobel Prize for their work on uncovering the structure of DNA. Sadly, Rosalind Franklin died in 1958 and did not live to receive an award. The discovery of the structure of DNA has given us a far greater understanding of genetics. Watson went on to head the Human Genome Project (see pages 20–21), which aimed to uncover all of the genes found in humans. This project was completed and published in 2003.

▲ This is a computer-generated drawing of the structure of DNA. The shape is called a double helix. DNA is invisible to the naked eye.

TIME TRAVEL: THE HUMAN GENOME PROJECT

In 1990, an international team of scientists from the United States, the United Kingdom (UK), France, Germany, China, and Japan set out to uncover the human genome. The human genome is the total of all the genetic information (genes) held in the cells of the human body. DNA molecules carry this information (see pages 18–19). The team aimed to explain what each part of the DNA molecule was responsible for. The results of the Human Genome Project were published in 2003, but the huge task of analyzing all of the information is still ongoing.

▲ This is the logo for the Human Genome Project.

RIVALRY

Before the 1990s, other teams of scientists had unraveled the genome for much simpler organisms, such as bacteria and yeast, but mapping the human genome was a far greater task. A private group claimed they would reveal the human genome within months, threatening the international project. They also claimed that once they had done this, they would charge anyone who wanted access to the information. Finally, both the private and international teams published the human genome in 2003. Fortunately, although both teams have continued to work in this field, the information is free of charge to its users.

▼ An IBM supercomputer, similar to this one, is used to analyze the human genome information.

WHY IS THE HUMAN GENOME SO IMPORTANT?

Since the structure of DNA was discovered (see pages 18–19), scientists have believed that certain traits and diseases are inherited through our genes. A full understanding of the human genome means that scientists are closer to knowing which genes are responsible for which traits and diseases. Therefore, if a disease is inherited, scientists may be able to pinpoint the gene responsible. For example, women with a certain type of breast cancer carry a particular mutated gene.

GENE THERAPY

Scientists have developed a technique called "gene therapy" to attempt to cure damaged and mutated genes to prevent the development of certain diseases. To date, this technique has not been very successful, but it is a fast-moving area of research. The ultimate goal is to use the human genome to discover new ways of treating, diagnosing, and perhaps one day preventing thousands of human disorders.

GENETIC TESTING

The genetic testing of embryos is one of many controversial issues connected with the Human Genome Project. In a genetic test, scientists scan a DNA sample for mutations that indicate that the embryo may develop a disease.

Imagine a situation in which a couple wants a child and is able to have its embryos genetically tested before they are implanted into the female for a pregnancy. If the genetic test shows that the embryo is not perfect in some way, should the couple be allowed to reject it? How far should people be allowed to "design" their own children? There are those who agree with embryonic genetic testing. They believe that if there is a way of curing otherwise incurable diseases, then why should we not do so?

Others believe that it is wrong to tamper with nature in this way and that we need limits on such testing. For example, some believe that it is acceptable to prevent illness, but that it is wrong to choose the eye or hair color of a child. This area of genetics is one that

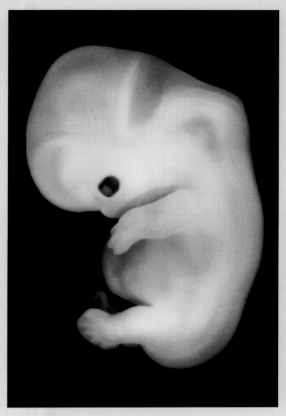

▲ How do you feel about genetic testing on human embryos?

will be debated for many years to come. However, a further problem is that the debate about genetic testing does not move as fast as the research itself.

A NEW VIEW OF LIFE

Scientists now have a "parts list" for how an organism operates, maintains, and reproduces itself. What is largely unknown is the function of most genes and how cells in the body use the information from the DNA to come alive. It is likely that future research will not look at individual genes, but rather at systems as a whole. Research on this larger scale will look at how genes behave in particular environments, how groups of genes act in tissues and organs, and how thousands of genes interact to create "life."

6

Natural selection

You may have heard the phrase "survival of the fittest." This means that only the healthiest organisms, or those best suited to their environment, live long enough to pass their characteristics on to the next generation. For example, some organisms have skin or fur that blends in well with their surroundings, so predators cannot easily see them. Others come out only at night when there is less danger of being preyed upon. Clearly, these organisms have characteristics or lifestyles that help them beat their competitors in the survival game. If they live long enough to reproduce, they will pass on these characteristics to their descendants. We call this trend **natural selection**.

NATURAL SELECTION OF THE PEPPERED MOTH

A famous example of natural selection occurred in the UK in the 1800s. There are two varieties of peppered moth; one is dark gray, and the other is paler and speckled. Both varieties of peppered moth feed at night and are preyed upon by birds during the day. To avoid being eaten in the daytime, the moths rest on tree trunks, which provide them with camouflage.

In the early 1800s, the speckled moth population was large, but dark moths were rare. The dark form could be easily seen on tree trunks and made excellent prey for birds. However, during the 1800s, many forests were cut down, and factories were built in their place. The factories churned out large quantities of soot, which discolored buildings and plant life. The discolored buildings and trees provided new camouflage for the dark moth, as it could "hide" itself during the day.

By the end of the 1800s, the dark moth population had grown massively, and speckled moths had become quite rare. The dark moths were living long enough to reproduce and pass on their genes to their offspring. At the same time, the speckled moths suffered greater predation and often did not survive long enough to reproduce.

◀ The dark moth is well camouflaged on this soot-colored tree trunk. The pale moth is easily seen and eaten by predators.

NATURAL SELECTION IN PEACOCKS

Male peacocks have huge, brightly colored tail feathers, which are used to impress and attract a mate. The tail feathers also make it difficult for the peacock to escape predators. If the male is able to live long enough to mate and produce offspring, he is likely to be a very strong specimen. In this way, the genes that indicate his "fitness" for survival and his brightly colored tail feathers are passed on. This type of natural selection is called sexual selection.

▶ This peacock is indicating he is a strong individual.

NATURAL SELECTION IN PLANTS

Many plant species have colored petals and sweet scents to attract insects. These features are present as a result of natural selection. The plants that attract the most insects spread their pollen (sex cells) to a large number of other plants. The offspring of such plants will also carry the genes that describe bright petals and a sweet scent. Therefore, the trait is selected and conserved in future generations. There are also countless numbers of species that have been unsuccessful in their adaptation. Such organisms are now **extinct** (see pages 38–41).

◀ This flower ensures the survival of its species by attracting insects.

INVESTIGATE

▶ Find out which features or abilities the following animals have developed as a result of natural selection to ensure the long-term survival of their species: (1) lions and (2) chameleons.

CHARLES DARWIN

Charles Darwin (1809–1882) was a British naturalist who became famous for proposing the theory of **evolution** by natural selection (see pages 22–23). Darwin studied animal and plant life across the globe, particularly in the Galapagos Islands. He came to the conclusion that all life on Earth evolved (developed gradually) over millions of years from a few common ancestors by the process of natural selection. But what led him to believe this?

DARWIN'S FIVE-YEAR VOYAGE

In 1831, Darwin sailed the HMS *Beagle* on an around-the-world voyage. During his five-year journey, Darwin spent time studying wildlife on the Galapagos Islands, which are located 600 miles (960 km) off the coast of Ecuador, South America. He observed, collected, recorded, and described hundreds of fossils, rocks, fish, insects, plants, birds, and mammals throughout the journey. He also sent many samples back home to England.

IMPORTANT DISCOVERIES

Upon Darwin's return in 1836, an ornithologist (bird scientist) named John Gould at the London Zoo Museum took up the study of various bird samples Darwin had sent him. Darwin thought the birds were entirely different species—wrens, blackbirds, and finches—but Gould discovered that they were all distinct species of finches.

Darwin studied the finches further and realized that they had obvious differences depending on the island from which they came. The finches from some of the Galapagos Islands had curved beaks, which were ideal for feeding on vegetarian matter. Finches from other Galapagos Islands had straight beaks, ideal for pecking at insects on tree bark.

▼ The Galapagos Islands are home to many weird and wonderful creatures, such as these iguanas.

DARWIN'S THEORY

For the next 20 years, Darwin thought long and hard about the finches and all of the other creatures he had studied and observed. He eventually came to the conclusion that all species were on Earth through a gradual change over millions of years by the process of natural selection. For this to take place, three things were needed:

(1) The organisms had to be able to reproduce.
(2) There had to be a source of variation (see pages 6–13). For example, a genetic mutation was the likely original cause of the change in beak shape in Darwin's finches.
(3) "Survival of the fittest" had to occur, which means that the best individuals reproduced. The finches with the best beak shape for getting food on their particular island were the "fittest" and lived long enough to reproduce.

Eventually, the different populations of finches changed so much that they became new species. This is known as **speciation**.

DARWIN'S BOOK

Alfred Wallace, a young British naturalist, collected animals in Malaysia where he lived and had drawn the same conclusions as Darwin. Once Darwin realized this, he worked long hours to write his book and publish his ideas on natural selection and evolution. Finally, in 1859, some 23 years after returning from his voyage, he published his famous book *The Origin of Species by Means of Natural Selection*.

◀ These finch species developed different beak shapes depending on the available food sources.

OUTRAGE

Darwin published his ideas in Britain at a time when the Catholic Church was very powerful and influential. What Darwin proposed caused an outrage because it conflicted with the belief that God created all creatures. Darwin risked becoming a social outcast. Today, the conflict between evolutionary ideas and religion often centers on "creationism" or "intelligent design." People who believe the Bible is literally true find that Darwin's ideas do not agree with their religious views and maintain instead that all species were created at once. However, almost 150 years since the publication of his life's work, Darwin's theories are widely accepted by the scientific community.

▼ The beak of this woodpecker finch is perfectly suited to hold a stick, which it uses to extract insects from the tree bark.

7

Artificial selection

The process of natural selection results in the survival of those organisms with the "best" features. This allows them to reproduce and pass their desirable genes to their offspring. However, humans often control the breeding of animals and plants to obtain particular variations in the offspring. This is called **artificial selection**.

ARTIFICIAL SELECTION IN PLANTS

Gardeners and farmers often look for plants or crops that will produce a good yield or flower for a long period of time. This is not easy, particularly in the winter when daylight hours are reduced and temperatures are low. Plants that grow every year go into a type of hibernation during the winter. Gardeners can prolong the yearly "lifetime" of their plants by putting them into a greenhouse, where conditions can be controlled. However, this is not always practical when there are many plants or when the plants cannot be easily moved. So, growers use artificial selection to create plants that will produce excellent yields or flower for a long time period.

HOW IS THIS DONE?

The winter-flowering pansy can be grown throughout the winter. It is frost-resistant and produces flowers in the coldest months of the year. To create such plants, growers choose long-flowering specimens from a normal population and breed them together. Offspring from this cross are observed, and those that flower the longest during the winter are selected. Seeds collected from those plants are germinated, and another generation is grown. The process is repeated until the desirable feature has become strong enough in the plant.

This kind of artificial selection procedure can be used to select a wide variety of different characteristics, depending on the requirements of the gardener or farmer. For example, artificial selection has been used to generate frost-resistant wheat called Marquis wheat. This crop can be grown at colder times of the year than other varieties, which increases its yield.

◀ Winter-flowering pansies produce flowers in the coldest months of the year, between October and May.

PROBLEMS WITH ARTIFICIAL SELECTION IN PLANTS

Artificial selection sounds like a harmless process and provides us with plants that are useful for our everyday lives. However, there is one major drawback. As each generation is produced, only a small number of plants are crossed to create the next generation. In this way, the desirable feature is selected, but many plants with other good genes are discarded. The plants in each generation become more and more alike, and there is a narrowing of what we call the "**gene pool**."

This can be a big problem when an environmental factor, such as disease or insect infestation, comes along. Because all of the plants are genetically very similar, if one plant dies, then there is a very high possibility that all of the plants will die. If the gene pool is wider, there are likely to be some plants capable of surviving.

ARTIFICIAL SELECTION IN ANIMALS

Artificial selection is used in the animal world, too, and this plays an important role in

producing cattle and sheep for their most desirable characteristics. Breeders aim to produce cattle for their milk or meat and sheep for their fleece (above).

HOW DOES IT WORK?

The cattle that produce the best milk or meat are bred with each other. This cross can be carried out naturally, by putting a bull into a field of female cows, or by artificial insemination. This involves the farmers using the sperm (sex cells) of a prize bull to fertilize his cows.

▲ Prize bulls are chosen for desirable characteristics such as their size or strength.

The offspring in this generation are selected for good milk or meat production and are crossed again. The process is much the same as it is in plants, except animals usually take longer to reach maturity, and the results of selective crosses take longer to be seen.

Artificial selection creates the same problems in animals as it does in plants. The gene pool narrows, and the selected animals may be at risk of disease or disability. Pedigree dogs often suffer from inbred traits. The bulldog is bred for its wrinkled appearance and wins prizes for this look. However, the wrinkles lead to breathing problems. In addition, intensively reared animals often suffer from diseases and illnesses that spread rapidly through the population as a result of poor conditions and treatment.

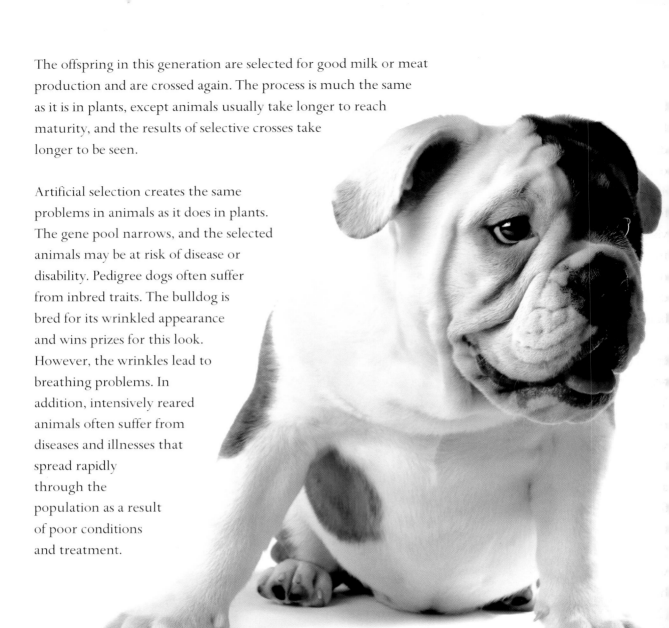

▲ Dogs with wrinkled faces often have breathing problems.

DID YOU KNOW?

▶ In his book *In the Language of Genes*, Steve Jones wrote, "There is little doubt that the most important event in recent human evolution was the invention of the bicycle." But what difference does owning a bicycle make? Before transportation systems were invented, no one traveled very far, and marriages were usually between people who lived within a few miles of one another. With the invention of transportation systems, couples formed between people who lived in different towns, cities, and countries. As a result, genetic mixing happens far more today than it did in the past.

ARTIFICIALLY SELECTED FOX

A farm in Siberia, Russia, has bred a domesticated fox, which looks and behaves just like a pet dog. This selective breeding occurred over a period of just 45 years. Scientists say that the foxes have the ability to follow visual expressions and movements of the hand in much the same way as a domestic dog. They call this behavior "social intelligence."

HOW HAS THE DOMESTIC FOX BEEN BRED?

The domestic foxes are the result of crosses between 100 vixens (females) and 30 male foxes. The foxes, originally from a fur farm, were crossed to select for "tameness." Tests for tameness included:

(1) Seeing whether the foxes would choose to make contact with a human's hand at feeding time.

(2) Seeing whether or not they would bite or run from a human.

After several generations, the scientists noticed that some of the new foxes were distinctly different from the wild foxes. The new foxes were

▲ These baby domestic foxes look a lot like dogs.

eager to receive attention from the human handlers and even whimpered, barked, and wagged their tails to attract attention. There were also physical changes. Their coats developed white patches like those of dogs, their snouts became shorter, and their ears became more floppy. What is most interesting about this domestication is that it shows us that while we are selecting for one feature, we can at the same time unintentionally select for another feature thought to be completely unrelated. Scientists also hope that the study of these foxes may give clues to the development of social intelligence in humans and other animals.

TEST YOURSELF

▶ Think of some other examples of selective breeding in the plant or animal kingdoms. What is being selected for and why?
▶ For one of these examples, explain how the actual selection could be done.

Time travel: Stem cells and cloning

Scientists have developed an advanced method of artificial selective breeding called **cloning**. This produces individuals that are genetically identical to the "parent" individual. It is currently illegal to create human clones in most countries, but it is possible to create cloned embryos from which **stem cells** can be taken.

How does cloning work?

(1) An egg is taken from a donor woman, and the nucleus is removed.

Donor egg has its nucleus removed.

The donor egg without its nucleus.

(2) An ordinary cell is taken from a second donor, who can be either male or female. The nucleus is removed.

(3) The nucleus from the donor cell in step 2 is placed into the egg cell from step 1. This nucleus contains all of the genetic material needed to create the new individual (the discarded egg nucleus only contains half of the necessary genetic material).

The donor nucleus is placed into the empty egg cell.

(4) The egg cell with the donated nucleus is given a short electrical impulse, which causes it to begin dividing to form an embryo.

(5) In theory, the embryo can be implanted into a surrogate mother and allowed to come to full term to produce a cloned human being. This procedure is not permitted in most countries. Research scientists are allowed to let the embryo divide and grow for only up to two weeks. During this time, valuable stem cells are produced.

The egg begins to divide and grow.

What are stem cells?

Stem cells are unspecialized cells. They have the remarkable ability to develop into any cell type. Stem cells are present in early human embryos and are very valuable to research scientists. When a stem cell divides, each new cell can either remain as a stem cell or become a more specialized type of cell, such as a skin or muscle cell.

Advantages and disadvantages

There are many diseases that can affect an individual's quality of life. Parkinson's disease occurs as a result of damage to the nervous tissue. In theory, a patient could donate a nucleus to create a cloned embryo. The stem cells from the cloned embryo could be removed and placed back into the original donor, where they could be made to grow into new nervous tissue. Scientists believe

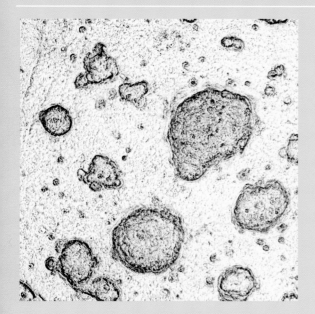

▲ These are stem cells. They are each around 0.0004 inches (0.01 mm) in diameter.

that, although a long way off, stem cells could be use to treat a great many diseases. The clear disadvantage of stem cell research is that a new life is created and then destroyed. Many people believe that the destruction of human life at any level of development is wrong.

AN ALTERNATIVE?

Stem cells have also been found in some parts of the human body, such as bone marrow, brain, and liver tissue. There are very few stem cells in each type of tissue, but scientists are looking into ways of growing them in the lab. Whether adult stem cells will be a useful medical tool remains to be seen, but early studies are promising.

Heart failure is measured using a quantity called "ejection fraction." The ejection fraction is the amount of blood released by the heart with each beat. In healthy people, this quantity is 55 percent, but in those with severe heart failure, it is less than 35 percent. In a recent study, adult stem cells were taken from the patients' own hip bones and injected into the damaged heart tissue. After six months, patients who had received stem cells had increased the ejection fraction to 46 percent. The stem cells appeared to have specialized and become heart muscle cells. Although in early stages of investigation, it seems that adult stem cells could be used to repair damaged tissues.

▼ This is synthetic bone. Stem cells are put inside it, where they grow into new bone tissue. The new bone is used to repair damaged bones in the human body.

8

Evolution uncovered

Charles Darwin's theory states that individuals of a species who are best adapted to their environment will survive long enough to breed and pass on their favorable genes to the next generation. This means that only the fittest will survive, and species will gradually change. Some change so much that they form new species. This is the process of evolution by natural selection. Evolution is a gradual process that can take millions of years. It is supported by a huge amount of powerful evidence.

▲ These fossilized fish lived millions of years ago.

FOSSIL EVIDENCE

Fossils are the preserved remains of plants or animals found in rocks. When animals and plants die, they become buried under the ground. Layers of soil and rock build up on top of them, and as this happens, the pressure rises. The lower layers are compressed into rocks, which can perfectly preserve the plant and animal remains into fossils. Rocks can be aged using fairly simple techniques, and this tells us when the plants and animals found within them lived. Dinosaur skeletons, footprints, and even an ichthyosaur (dinosaur species) embryo inside its mother have been discovered.

One of the most stunning examples of fossilized evolutionary evidence is that of the horse. Fossils tell us that the modern horse evolved from an earlier species called eohippus, which lived around 60 million years ago. By studying fossilized bones and teeth, we have learned that the four-toed eohippus ate leaves from low-hanging branches in forests. Over millions of years, eohippus evolved into mesohippus, which had three toes and ate leaves and grass. Pliohippus was the next horse to evolve and had a single-toed hoof and ate mostly grass. Eventually, modern horse, or equus, evolved. Equus eats grass and has a single-toed hoof. It has longer legs and can run faster than its ancestors.

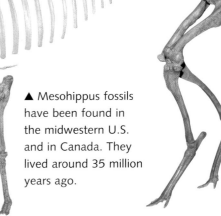

▲ Mesohippus fossils have been found in the midwestern U.S. and in Canada. They lived around 35 million years ago.

GEOGRAPHICAL EVIDENCE

Similar animals and plants are found in similar environments wherever they are in the world. Darwin visited the island of Tristan de Cunha, which lies between Africa and South America. On this island, he found plants with features from species found in Africa and South America. Darwin concluded that the island plants were the result of seeds that had been carried by the wind or water from the mainland to the island. Once there, they evolved by natural selection to occupy the specific habitat of the island.

Similarly, rhinoceroses are found on two separate continents—Africa and Asia. It is unlikely that they each formed spontaneously. Instead, fossil records found in different parts of the world show us that today's rhinos are descended from a single common ancestor that lived more than 30 million years ago. When the evolution of the ancestor diverged (separated), one line led to the one-horned rhino of India, and the other led to the two-horned rhino of Africa.

▲ The one-horned Asian rhinoceros

▲ The two-horned African rhinoceros

COMPARATIVE ANATOMY

Scientists study and compare parts of animals, such as the heart or feet, to see whether there are any similarities that would indicate a common ancestor. The structures of the hearts of amphibians, reptiles, and mammals demonstrate a clear pattern of evolution. The three-chambered amphibian heart has a chamber (1) where all of the blood flowing to and from the heart can mix.

The reptile heart has an incomplete partition, which allows a little mixing. The mammalian heart has a wall that completely separates the heart. This is strong evidence that some amphibians evolved into reptiles and that some reptiles evolved into mammals. Clearly, a large number of amphibians and reptiles did not evolve further, or they wouldn't exist today.

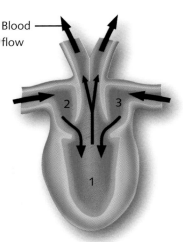

▲ An amphibian's three-chambered heart.

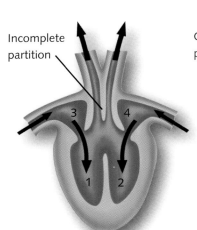

▲ A reptile's four-chambered heart with incompletely separated chambers.

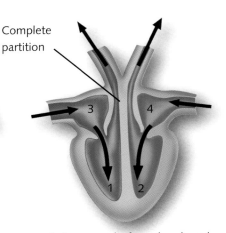

▲ A mammal's four-chambered heart with completely separated chambers.

EMBRYOLOGY

Embryology is the study of organisms and their development from a fertilized egg until they are ready for birth or hatching. It can provide important evolutionary clues. Similarities between species are often seen in embryos but not in the adult form. Embryos of different species that look similar indicate that they may have had a common ancestor. The embryos of a chicken, a pig, and a human look strikingly similar. Among other features, they all have a head with gill pouches, small limb buds, and a tail-like structure.

▼ A pig embryo

▼ A human embryo

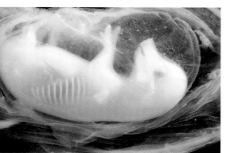

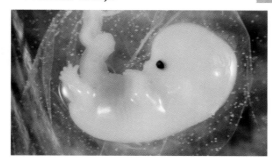

▲ A chicken embryo

MOLECULAR EVIDENCE

Almost every organism on Earth contains DNA. DNA is not only the genetic code of an organism; it is also a record of its evolutionary history. By examining DNA molecules, scientists can discover how closely organisms are related. The number of differences in the DNA between two species is directly linked to the length of time since they shared a common ancestor.

Pseudogenes (false genes) are genes that no longer describe a specific function or trait. These are carried in the DNA anyway and are a very useful tool for finding out how closely two species are related. Some genes exist in two animals because the animals have similar habitats and lifestyles, such as zebras and horses. However, if the same pseudogenes are found in zebras and horses, the only logical explanation is that they have a common ancestor.

The molecular evidence for evolution is overwhelming and in some cases gives us a better description of evolution than other sources of evidence. For example, scientists have long believed that whales are descended from land mammals. They thought that whales were

DID YOU KNOW?

▶ Some scientists believe that three different human species—*Homo sapiens* (modern humans), *Homo neanderthalensis*, and *Homo erectus*—coexisted on Earth for about 100,000 years. Traditionally, archaeologists thought that only one human species could exist at a time. In recent years, this belief has been shattered. Fossils prove that several human species occupied Earth at the same time hundreds of thousands of years ago. Today, only modern humans are thought to exist.

related to hoofed creatures, such as cattle, sheep, goats, and camels. By studying whales on a molecular level, scientists discovered that their genes that code for milk proteins are very similar to those of the hippopotamus, a hoofed animal.

In this case, molecular evidence has enhanced the fossil record and has told us that whales and hippos have a common ancestor.

▼ Hippos and blue whales share a common ancestor.

RECENT EVOLUTION

Bacteria exist in all corners of Earth. Some of them are harmless, and others cause disease. Since the introduction of antibiotics in the mid-1900s, bacteria have evolved strains that are antibiotic-resistant. Bacteria change rapidly as a result of mutations and natural selection, and in just one generation can change from being killed by antibiotics to being resistant. This has led scientists to develop many new antibiotics in the hope of staying one step ahead of the bacteria. Equally, the influenza virus evolves each year through mutations and natural selection, meaning that scientists constantly have to create new vaccinations.

TIME TRAVEL: THE HOBBIT OF INDONESIA

Our knowledge of evolution changes all the time as more and more evidence is uncovered. In September 2004, a group of Australian and Indonesian researchers discovered the bones of a tiny human-like creature buried in a limestone cave on the Indonesian island of Flores. It has been nicknamed the "hobbit," as it was only three feet (0.9 m) tall. This astonishing and controversial discovery has rocked scientists and their understanding of human evolution.

WHAT IS IT?

It was originally believed that the bones belonged to a child. However, investigation of the teeth shows that they are worn down and therefore belonged to an adult. The pelvis indicates that the individual was female, and the leg bones suggest that she walked upright. It is thought that the hobbit is not a *Homo sapiens* (a modern-day human), as it has very long arms and a small head the size of a grapefruit. Some of her head features represent those of a human species called *Homo erectus*, thought to have lived on a nearby island around 1.5 million years ago. However, her tiny size, similar to that of a three-year-old child, suggests that she is our first glimpse of a new species of human being.

Scientists have named the new species *Homo floresiensis*. Amazingly, *Homo floresiensis* is thought to have lived as recently as 18,000 years ago, which in evolutionary terms is practically yesterday. Since the original find, bones and teeth from six other individuals have been discovered.

▶ **This is the cave in the which the "hobbit" was found on the island of Flores.**

SIZE AND TOOLS

Homo floresiensis' tiny size may be a result of being confined on an isolated island with few food resources. In addition, size and strength are less important on an island where there are few large predators. On Flores, the only large predator was the Komodo dragon.

▼ *Homo floresiensis'* **skull (top) compared with a modern-day human's skull (bottom).**

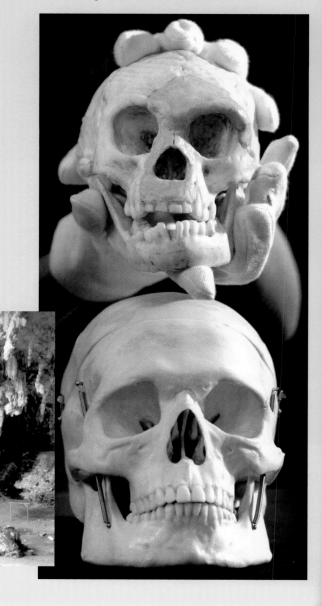

Scientists hope to unravel the mystery of *Homo floresiensis* and in particular uncover the origins of the most baffling characteristic—its small brain size. Sophisticated tools have been unearthed that seem far beyond the scope of *Homo floresiensis'* tiny brain. Scientists are still studying the bones and hope to extract DNA to help solve the puzzle and place *Homo floresiensis* in evolutionary history.

▼ An artist's impression of *Homo floresiensis*.

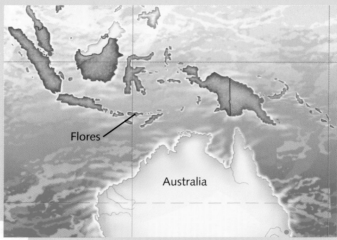

▲ The island of Flores is in Indonesia.

CONTROVERSY

There are those who doubt whether the bones represent a new human species. Certain medical conditions cause stunted growth in *Homo sapiens*, and some scientists have suggested that this accounts for the tiny body size and adult teeth. Other researchers think that it could be a diseased form of *Homo erectus*, or an individual that was deprived of food. However, since more than one specimen has been discovered, these theories seem unlikely. Where this species came from, and how it got to Flores, remains a mystery and will be a topic of research for many years to come.

LEGENDARY TALES?

Today's inhabitants of the island of Flores tell folk stories of little people called Ebu Gogo, which translates as "the grandmother who eats anything." They describe Ebu Gogo as being three feet (0.9 m) tall, hairy, and prone to communicate either through murmuring or through repeating parrot-fashion what others have said. Some scientists speculate that these legends are based on truth and have even suggested that the little species is still in existence. Whether or not *Homo floresiensis* is proved to be a unique species that coexists with us today, evolutionists are slowly unraveling the story of human evolution.

9

Extinction

When a whole species dies out, we say that it has become extinct. Extinction is usually a natural process, and it is estimated that 99.9 percent of all species that have ever lived are now extinct. For many species, the process of extinction takes so long that new species evolve before the original species disappears. However, extinction can also occur as a direct result of man's actions, and today there are many precious creatures on the brink of being wiped out.

NATURAL EXTINCTION

In nature, species become extinct when they reach the end of their evolutionary period on Earth. The length of this period depends on how well a species adjusts to changes in climate, habitat, predators, and food sources. Natural events, such as volcanic eruptions, tsunamis, and sudden climate change, can cause mass extinctions if the event is on a large enough scale.

There have been several mass extinctions in Earth's history. Around 440 million years ago, more than 75 percent of all species became extinct. Primitive sea creatures were the worst affected. The cause of this extinction is thought to have been a dramatic change in climate—an **ice age**—during which Earth and everything on it froze for thousands of years. The last mass extinction occurred 65 million years ago, but smaller waves of extinctions have happened more recently. Just 11,000 years ago, two important events took place. The ice sheets from the last ice age retreated, and prehistoric man entered North America from Siberia. We know that at this time, more than two-thirds of North America's mammals disappeared, including the short-faced bear and the saber-toothed cat.

MAN-MADE EXTINCTION

Today, one of the biggest threats to wildlife is man. Throughout history, humans have hunted animals for food, clothing, souvenirs, and trophies. Birds with beautiful displays of brightly colored feathers, such as parrots and birds of paradise, were killed to make attractive headdresses. Unfortunately, the birds were often killed more quickly than they could be replaced, which caused the permanent loss of many species.

▲ Some birds are hunted for their colorful feathers.

Man has also hunted animals for more obscure features, such as elephant ivory, snakeskin, gorilla paws, and rhinoceros horns. Each of these items is meant to bring buyers good luck or provide them with a unique ornament to display. Some people also like to have shoes or handbags made from rare reptile skins. Importing all of these items is illegal in many countries. Man has also destroyed the habitats of many species. Large areas of the Brazilian rain forests have been cleared for farming and to obtain timber for building. This has led to the loss of untold numbers of plant and animal species. It is thought that half of the world's tropical rain forests have been cut down since 1985. The tragedy of such destruction is that these rain forests contain more than half of the world's endangered species.

▲ The number of African elephants fell from 1.2 million in the 1970s to around 500,000 today, mainly as a result of the ivory trade.

ENDANGERED SPECIES

Some species, while not yet extinct, are in serious danger of becoming so. When activities such as hunting and habitat destruction reach a critical level, species numbers become so low that they are in danger of being made extinct—they are endangered. Currently, in western and central Africa, forest animals have been hunted for food to such an extent that many are now threatened with extinction. Approximately one million tons (1 million t) of forest animals are hunted and killed in these regions every year. One threatened species is the bonobo, which is believed to be man's closest relative.

▲ Bonobos are the rarest of all the great ape species.

DID YOU KNOW?

▶ The tragic tsunami in Southeast Asia in December 2004 not only killed and displaced tens of thousands of people, but it may also lead to extinctions. Many habitats, including mangroves, coral reefs, wetlands, and forests were damaged, and freshwater lakes were poisoned by salt water.

TIME TRAVEL: DINOSAUR EXTINCTION

Dinosaurs dominated Earth for 165 million years. They ranged in size from huge, 50-foot-tall (15 m) creatures to small, chicken-sized animals. Around 65 million years ago, there was a catastrophic event that killed off approximately 75 percent of life on our planet; most famously, the dinosaurs were wiped out.

HOW DID THE DINOSAURS BECOME EXTINCT?

The most widely accepted theory for dinosaur extinction is that an asteroid slammed into Earth. Evidence supporting this theory comes from studies of iridium levels in rocks. Iridium is a precious metal found in small amounts in Earth's rocks but in much higher concentrations in asteroids. Scientists have found a very thin layer of iridium in rocks worldwide, with a chemical makeup similar to that of an asteroid. An asteroid–Earth collision could have caused iridium to disperse across the globe and settle, forming this thin layer.

THE ASTEROID THEORY EXPLAINED:

(1) An asteroid from outer space crashed into Earth. This impact may have been in the Yucatan region of Mexico, where there is a crater approximately 125 miles (200 km) wide—large enough to have caused a global catastrophe. The explosion this caused would have been 10,000 times greater than all of Earth's nuclear weapons going off at once.

(2) The impact caused massive earthquakes and tsunamis, resulting in widespread destruction at ground level.

Crater site

▲ The impact crater is half on land and half in the sea in the Yucatan region of Mexico.

(3) Hot debris was sent flying in all directions, leading to forest fires and the loss of many species and their habitats.

(4) Millions of tons of dust and debris were thrown into the atmosphere, which began to settle over a very large area.

(5) Acidic gases were released from Earth's damaged crust. These entered the atmosphere, circulated the world, and eventually fell as acid rain, seriously polluting waterways.

(6) Dust and debris that did not settle blocked light and heat from the sun. Earth's climate quickly cooled, and a perishing winter set in, lasting many months. Large numbers of species could not

◀ An asteroid is thought to have wiped out the dinosaurs.

survive such a sudden change in climate. (7) Following the "winter," a buildup of carbon dioxide in the atmosphere prevented any heat left on Earth from leaving. This greenhouse effect had devastating effects on global weather patterns, resulting in the extinction of yet more species.

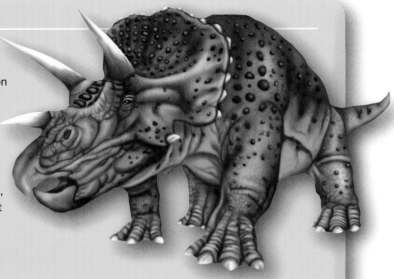

THE CONSEQUENCES

All dinosaurs, whether they lived on land, in the sea, or in the air, were made extinct. The lack of sunlight would have caused the death of most plant species. Herbivorous (plant-eating) dinosaurs would have starved and died. Carnivorous (meat-eating) dinosaurs fed on herbivorous dinosaurs and would have also starved. Dinosaurs needed sunlight to keep warm—a further factor making their survival impossible. The survivors were mostly small mammals, birds, and fish that were able to adapt to the new conditions on Earth. As with all mass extinctions, there was a sudden evolutionary burst of new species to fill Earth's new habitats.

MEGA-VOLCANOES

Not everyone accepts the asteroid theory. Around 65 million years ago, there was extreme volcanic activity across the world. Volcanoes spewed dust and acidic chemicals into the atmosphere, and this may have changed Earth's climate enough to trigger a mass extinction. The only way we could ever know for certain is if we could travel back in time.

COULD IT HAPPEN AGAIN?

Mass extinction events have occurred regularly throughout Earth's history. Researchers estimate that they occur every 26 to 30 million years, possibly as a result of collisions with a cloud of comets, called the Oort cloud, crossing Earth's path. While a mass extinction event is highly likely, we are pretty safe for the time being, or at least the next 10 million years.

Classification

As a result of evolution, there are millions of species living on Earth today. Our planet supports all of these species, and scientists are finding new ones all the time. It is easy to tell the difference between some species, such as between an oak tree and a wheat plant, whereas others look very similar, such as African and Indian elephants. A system of classification allows biologists to name all species.

FIVE KINGDOMS

There are five main kingdoms: animals, plants, prokaryotes, protists, and fungi. Each kingdom is split into several phyla. Each phylum group contains organisms that have features in common.

THE ANIMAL KINGDOM

The animal kingdom is divided into invertebrates (organisms without a backbone) and vertebrates (organisms with a backbone).

▲▼ This crab (above) and this beetle (below) are invertebrates. They have no backbone.

INVERTEBRATES

PHYLUM	FEATURES
Cnidarians	Sac-like body with tentacles (for example, a jellyfish)
Flatworms	Flat body with a mouth at one end
Roundworms	Long, thread-like body
Segmented worms	Body divided into segments (for example, an earthworm)
Mollusks	Carry a shell and have a muscular "foot" (for example, a snail)
Echinoderms	Contain five parts and have spiny skins (for example, a sea urchin)
Arthropods—crustaceans	Chalk-like exoskeleton (for example, a crab)
Arthropods—myriapods	Segmented bodies and many pairs of legs (for example, a millipede)
Arthropods—insects	Body in three parts; three pairs of legs and two pairs of wings (for example, a beetle)
Arthropods—arachnids	Body in two parts; four pairs of legs (for example, a spider)

VERTEBRATES

▶ This puffin is in the animal kingdom, is a vertebrate, and belongs to the bird phylum.

PHYLUM	FEATURES
Birds	Have feathers, wings, and lungs; lay eggs
Fish	Live in water; have scaly skin, fins, and gills
Amphibians	Moist skin and live on land and in water; lay eggs in water (for example, a frog)
Reptiles	Dry, scaly skin; breed on land; lay eggs with shells (for example, a lizard)
Mammals	Have hair or fur; have lungs; young grow inside the mother, who has mammary glands on which young can suckle

THE PLANT KINGDOM

▼ Moss

PHYLUM	FEATURES
Ferns	Have strong stems and roots and spores on the underside of the leaves
Conifers	Keep their needle-like leaves throughout the winter
Flowering plants	Produce flowers that contain the reproductive organs and produce seeds
Mosses and liverworts	Live in damp places and do not have a true stem or roots

THE PROKARYOTE, PROTIST, AND FUNGI KINGDOMS

KINGDOM	FEATURES
Prokaryotes	No true nucleus and can be seen only with a microscope (for example, bacteria)
Protists	Have a nucleus; can be single- or multicelled; live in water (for example, algae) or inside other organisms
Fungi	Made up of thin threads called hyphae; produce spores (for example, a mushroom)

▲ These mushrooms are in the fungi kingdom.

TEST YOURSELF

▶ For each of the following organisms, decide which kingdom and, if possible, which phylum, it belongs to:
Apple trees

Crocodiles
Eagles
Sea urchins
Bread mold

CLASSIFYING ORGANISMS

Phyla are divided into smaller groups, which themselves have subcategories. This system is used by biologists to classify any living organism. Each of the smaller divisions is shown in the table below for two species of rats. Both are common rats; however, the black rat is famous for transmitting the bubonic plague, which killed millions of people in Europe during the 1300s. The brown rat carries other diseases, such as salmonella.

The black and the brown rat belong to the same kingdom, phylum, class, order, family, and genus. They share many features. They each have mammalian characteristics such as fur and mammary glands. Both species have rodent characteristics such as a gestation period of roughly 22 days and the ability to produce between 4 and 7 litters in one year.

▲ Brown rats

However, there are subtle differences between these two rats, which means that they are different species and cannot breed together to produce fertile offspring. The black rat has slate black fur, whereas the brown rat is red-brown. The black rat has pointed ears and a long nose, whereas the ears of the brown rat are more rounded, and the nose is a little shorter.

▼ Black rat

	BLACK RAT	**BROWN RAT**
KINGDOM	Animal	Animal
PHYLUM	Vertebrate	Vertebrate
CLASS	Mammalia	Mammalia
ORDER	Rodentia	Rodentia
FAMILY	Muridae	Muridae
GENUS	*Rattus*	*Rattus*
SPECIES	*rattus*	*norvegicus*

DID YOU KNOW?

▶ The system of naming was introduced by a Swedish biologist named Carolus Linnaeus in his book called *Systemae Naturae* in which he gave all living organisms a two-part name that was based on their features. Human beings are named *Homo sapiens*.

TAXONOMY AND KEYS

Biologists who study classification are called **taxonomists**, and they identify species by asking questions. For example, if you wanted to identify whether an organism is a prokaryote, you could ask the following question:

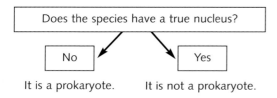

This diagram is called a key and is composed of only one simple question. However, keys can be extended to include a whole series of questions, all with "yes" or "no" answers. Imagine we have five animals that need to be identified—a snail, mouse, frog, jellyfish, and earthworm. The following key could be used:

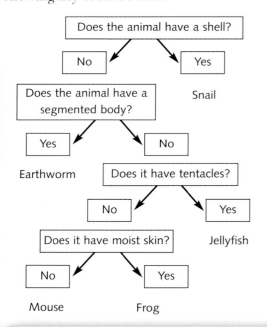

The questions do not need to be set out as a diagram. They can be listed, as shown below:

(1) Does the animal have a shell?
Yes . . . . it is a snail
No . . . . go to Q2

(2) Does the animal have a segmented body?
Yes . . . . it is an earthworm
No . . . . go to Q3

(3) Does the animal have tentacles?
Yes . . . . it is a jellyfish
No . . . . go to Q4

(4) Does the animal have moist skin?
Yes . . . . it is a frog
No . . . . it is a mouse

We do not have to memorize every detail about an organism in order to identify it. Instead, we use a key as a shortcut. The system of classification is continually being modified as new information and species are discovered. Organisms are moved from one genus to another as scientists make new observations. This improves our knowledge of the classification system and the living world.

▲ Keys are used to identify organisms.

TEST YOURSELF

▶ Choose five objects in your schoolbag and write a key to allow someone to identify them. Try your key out on your friends.

▶ Choose five of your friends and write a key to identify them.

Glossary

ALLELES – Variants of a gene. An allele is a member of a pair or series of genes that code for a particular feature, such as eye color or hair color.

ARTIFICIAL SELECTION – When humans deliberately reproduce plants or animals so that desirable traits are represented in later generations.

CELLS – The tiny building blocks that make up living organisms. Each cell contains a nucleus, which is the control center of the cell.

CHROMOSOMES – Threads of DNA found in a cell's nucleus.

CLONING – The creation of an individual who is genetically identical to the parent. Identical twins are clones, but clones can also be produced by genetic engineering.

CONTINUOUS VARIATION – Variation in characteristics that cannot be put into one category or another, but instead range from one extreme to the other. The changes are slight and overlap so much that they are thought of as being continuous, such as height or weight.

DISCONTINUOUS VARIATION – Variation in a characteristic that can be placed in one category or another, such as blood group.

EMBRYOLOGY – Study of the formation and development of organisms from a zygote, or fertilized egg.

EVOLUTION – The process by which all living things change slowly over millions of years. The change occurs because of slight variations in the genetic makeup that one generation passes on to the next.

EXTINCT – A species or group of species that has ceased to exist.

FOSSILS – The remains of living things, or their imprints, that have been left in rocks over a very long period of time.

GENE POOL – All of the genes in a population of organisms at any one time.

GENES – Specific segments of DNA on a chromosome that describe an individual's features.

ANSWERS

page 7 Investigate
Your graph should have shoe size on the horizontal axis and the number of people on the vertical axis. If five people had size seven shoes, for example, you would draw a horizontal line from "5" and a vertical line from "7." Where the two lines meet, you draw a cross. When you have done this for all of the different shoe sizes, draw a line, and connect all of the crosses. This should give you a normal distribution curve, which is a bell-shaped curve (right).

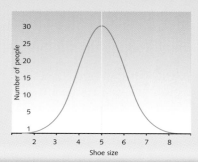

page 7 Test yourself
Weight and IQ are examples of continuous variation.
Gender, attached and free earlobes, eye color, and the ability or inability to hyperextend the thumb (hitchhiker's thumb) are all examples of discontinuous variation.

page 17 Investigate
The "tall" allele (T) is dominant over the "dwarf" (t) allele.
If you cross Tt and Tt, 75 percent of the offspring are tall plants, and 25 percent are dwarf plants.
If you cross TT and tt, 100 percent of the offspring are tall.
If you cross TT and Tt, 100 percent of the offspring are tall.

page 23 Investigate
To ensure the long-term continuation of their species:
(1) Lions have grown bigger and stronger and are able to run fast. Their pale fur is good camouflage against the grassy plains on which

GENOTYPE – The genetic makeup of an organism.

ICE AGE – A time period that lasts thousands of years, during which Earth is very cold. During an ice age, massive ice sheets cover much of Earth's northern hemisphere. The last ice age retreated around 10,000 years ago.

MUTATION – A change in the amount or arrangement of genetic material in a cell. The altering of the genetic makeup in this way can be advantageous, of no consequence, or fatal to the individual.

NATURAL SELECTION – The process in nature by which only the organisms best adapted to their environment survive and pass their genetic information on to the next generation.

NORMAL DISTRIBUTION CURVE – This is a bell-shaped curve on a graph, formed by plotting information that is normally distributed. This means that fewer individuals have extreme characteristics, such as a very large or very small foot size. Most individuals have characteristics that fall between the two extremes, such as medium-sized feet.

PHENOTYPE – The physical appearance of an individual that is determined by the genes and the environment. For example, blond hair is a phenotypic characteristic.

PSEUDOGENES – Naturally occurring genes that are altered in some way so that they do not have a function.

SPECIATION – The formation of a new species through evolution.

SPECIES – A group of organisms that can breed freely with one another, but not with members of other species. For example, cows are one species, and pigs are another. They cannot breed with each other to produce fertile offspring.

STEM CELLS – Unspecialized, immature cells that are capable of developing into any type of cell in the body. Stem cells are found in embryos and are called embryonic stem cells. Adult stem cells are also found in some parts of the adult body.

TAXONOMISTS – Scientists who deal with the identification, naming, and classification of living things.

---

they live. This allows them to remain undetected while they stalk and catch prey.

(2) Chameleons have developed the ability to change color. They do this to communicate with each other—for example, when they want to mate. Because this communication encourages mating, it could be viewed as a factor that helps the species survive.

page 29 Test yourself
Animal kingdom: Chickens can be selected for their ability to lay large eggs. This is done to increase profits made through the sale of eggs. How is it done? Chickens that lay the biggest eggs are crossed. The female offspring that produce the largest eggs are selected and bred again. The process is repeated until a large proportion of a generation produces large eggs.

Plant kingdom: Fruit trees are selected for dwarf varieties that can be

grown in pots. This makes them easy to transport and allows them to be grown in small gardens.

page 43 Test yourself
Apple trees = Plant kingdom, flowering plants
Crocodiles = Animal kingdom, reptiles
Eagle = Animal kingdom, birds
Sea urchin = Animal kingdom, echinoderms
Bread mold = Fungi kingdom

page 45 Test yourself
Example key for a pen, pencil, book, calculator, and key.
(1) Can you write with it? YES – Go to Q2. NO – Go to Q3.
(2) Is the writing gray? YES – It is a pencil. NO – It is a pen.
(3) Does it have pages in it? YES – It is a book. NO – Go to Q4.
(4) Does it have buttons? YES – It is a calculator. NO – It is a key.

Index

Page references in italics
represent pictures.

Photo Credits – *(abbv: r, right, l, left, t, top, m, middle, b, bottom)* **Cover background image** Q2A Creative **Front cover images** (bl) IFA Bilderteam GMBH/Oxford Scientific (tr) Iar Boddy/Science Photo Library **Back cover image** (inset) IFA Bilderteam GMBH/Oxford Scientific **p.1** (tr) www.istockphoto.com/Daniel Goodchild (bl) www.istockphoto.com/ Ingvald Kaldhussæter (br) www.istockphoto.com/Jeroen Peys **p.2** www.istockphoto.com/Ewa Brozek **p.3** (t) www.istockphoto.com/blaneyphoto (b) www.istockphoto.com/Ana Abejon **p.4** (tl) www.istockphoto.com/Jason van der Valk (tr) www.istockphoto.com/Bob Ainsworth (br) www.istockphoto.com/Steve McWilliam **p.5** www.istockphoto.com/ lamprey **p.6** Jose Luis Pelaez, Inc./Corbis **p.8** (t) Yann Arthus-Bertrand/Corbis (m) www.istockphoto.com/Fanelie Rosier (b) www.istockphoto.com/Stephanie Asher **p.9** (t) www.istockphoto.com/lamprey (b) www.istockphoto.com/Charleen Cole **p.10** (t) www.istockphoto.com/Ana Abejon (b) Andrew Harrington/naturepl.com **p.11** (t) www.istockphoto.com/Judi Ashlock (b) Philippe Clement/naturepl.com **p.12** Weill Rachel/Oxford Scientific **p.13** Ian Boddy/Science Photo Library **p.14** James King-Holmes/Science Photo Library **p.15** www.istockphoto.com/Ewa Brozek **p.16** (l) www.istockphoto.com/John Kerher (r) www.istockphoto.com/Louis Aguinaldo **p.17** Kit Houghton/Corbis **p.18** Science Photo Library **p.20** (t) US Department of Energy Genome Programs http://doegenomes.org (b) Louie Psihoyos/Corbis **p.21** Dr G Moscoso/Science Photo Library **p.22** David Fox/Oxford Scientific **p.23** (t) Ifa-Bilderteam Gmbh/Oxford Scientific (b) Gulin Darrell/Oxford Scientific **p.24** Doug Allan/Oxford Scientific **p.25** (t) Science Photo Library (b) Mary Plage/Oxford Scientific **p.26** www.istockphoto.com/Ingvald Kaldhussæter **p.27** www.istockphoto.com/Daniel Goodchild (b) Robert Dowling/Corbis **p.28** www.istockphoto.com/Stephanie Phillips **p.29** Irene Plyusnina/Current Biology **p.31** (t) www.istockphoto.com/Andrei Tchernov (b) Klaus Guldbrandsen/Science Photo Library **p.32** (t) www.istockphoto.com/Bob Ainsworth (b) Kevin Schafer/Corbis **p.33** (t) Mike Powles/Oxford Scientific (b) www.istockphoto.com/ Jeroen Peys **p.34** (l) Daniel Sambraus/Science Photo Library (m) Steve Allen/Science Photo Library (r) Oxford Scientific **p.35** www.istockphoto.com/Maciej Sekowski (b) Dan Shapiro/NOAA **p.36** (l) Reuters/Corbis (r) Rex Features **p.37** reproduced with the permission of Peter Schouten/National Geographic Society **p.38** www.istockphoto.com/Lisa F Young **p.39** (t) Reuters/Corbis (b) Martyn Colbeck/Oxford Scientific **p.40** (tr) NASA (tl) Courtesy NASA/JPL-Caltech **p.41** Zephyr/Science Photo Library **p.42** (t) www.istockphoto.com/Jason van der Valk (b) www.istockphoto.com/Eric Gagnon **p.43** (t) David Tipling/Oxford Scientific (m) www.istockphoto.com/David Freund (b) www.istockphoto.com/blaneyphoto **p.44** (t) Robin Redfern/Oxford Scientific (b) Bert & Babs Wells/Oxford Scientific **p.45** www.istockphoto.com/Steve McWilliam